Abstraction in Theory–

Zero Postulation Results

Theory of Everything

Book – 2

Subhajit Ganguly

V M A

Abstraction in Theory–Zero Postulation Results

V M A Publications

Science Books

ISBN-13: 978-1500216764
ISBN-10: 1500216763

First published in 2014

http://subhajitgangulyauthor.wordpress.com

DEDICATION

Dedicated to all humanity…

CONTENTS

ACKNOWLEDGMENTS

This book is a culmination of the good wishes of many individuals, without whom nothing would have been possible. I take this opportunity to thank them all from the bottom of my heart. Hope my endeavor does justice to all the good wishes and aspirations surrounding it.

1 Preface

Over some years now, a large part of the energies of the scientific community has been employed solely for finding a theory that will fit in all known happenings of the physical world. Various groups of scientists have tried to attack the problem from different ends. Some of these theories have been partly successful in explaining the known physical world. However none of these theories have been without shortcomings. Be it the much lauded String Theory or the Quantum Gravity postulation or any other such attempts towards arriving at a Theory of Everything, none have been proved to be foolproof. To say the least, nobody can deny that there is room for much improvement before we can even start thinking truly towards such a theory that would describe the known world satisfactorily and provide for a single basis of understanding the four forces in nature.

On top of that, we have the newly emerging problems of 'Dark Energy', 'Dark Matter' and the like. These realms are yet to be accepted by the scientific community officially, but nonetheless, they are most definitely at least a few parts of mysteries that remain unexplained. A good and effective Theory of Everything must aim towards explaining such mysteries too. Sadly, we have no theory as yet that fulfills these criteria.

From the dawn of civilization, human beings have tried to find out order in the chaotic world surrounding them. It has however never been easy to find a solution to explain a given system while being a part of that system. The best bet is to find out the most fundamental components within the system and building a theory round these. In other words, a theory that is able to describe the world in totality has to keep the number of basic postulates it depends upon to zero or near zero. Deductionism hits a dead end in this regard. On the other hand, abstraction as the starting point of building up a theory may be seen to be of fitting use. It would be much more than a new way of tackling the problem. Even abstract postulates do away with the shackles that bind our theories into the system and bar them from being total descriptions of the system. The abstraction we are talking about here may be defined as, "Postulation of non-postulation" or, in other words, "A system of postulation that gives equal weights to all possible solutions inside the system and favours none of such solutions over others."

Abstraction automatically gives rise to optimized solutions within the universal set of all possible solutions, as has been shown in this book. It is these optimized solutions that make up and drive the non-abstract parts of the world, while the non-optimized solutions remain 'hidden' from the material world, inside the abstract world.

Starting from a basis of no postulation, we build our theory. As we go on piling up possibilities, we come to a similar basis for understanding the four non-contact forces of nature known till date. The difference in ranges of these forces is explained from this basis in this book. Zero postulation or abstraction as the basis of theory synthesis allows us to explore even imaginary and chaotic non-favoured solutions as possibilities.

3

With no postulation as the fundamental basis, we are thus able to pile up postulated results or favoured results, but not the other way round. We keep describing such implications of abstraction in this book. We deal with the abstraction of observable parameters involved in a given system (quantum, relativistic, chaotic, non-chaotic)and formulate a similar basis of understanding them.

Scaling of observable parameters in adequate ways is shown to unite the understanding of worlds of the great vastness of the universe and the minuteness of the sub-atomic realm. Finally, the mysteries involving 'dark energy' and 'dark matter' are uncovered using such an approach.

This book is a culmination of the good wishes of many individuals, without whom nothing would have been possible. I take this opportunity to thank them all from the bottom of my heart. Hope my endeavour does justice to all the good wishes and aspirations surrounding it.

1 Abstraction and the Standard Model

Key Ideas:

We study the Standard Model in light of the Zero-Postulation of the Theory of Abstraction. Yukawa Coupling, chiral superfields, the SUSY model, Interacting Boson Models (IBMs), Clebsch-Gordan coefficients, Interacting Boson-Fermion Model (IBFM), etc., are some of the concepts that we study in this paper. Non-commutative geometry seems to come very handy in describing the quantum world. Bosons and fermions both seem to be governed by the rules of such geometry. The principle of conservation of boson number inside a system is seen to follow directly from the Abstraction Model. The IBMs are seen to obey the Laws of Physical Transaction that follows from Zero-Postulation. The chaotic superfields at the requisite scaling-ratio yields necessary equation-parameters needed to describe them at that given scaling-ratio. This is seen to be independent of the choice of scale, but at smaller scaling-ratios, we have less loss of information. At a higher scale, we seem to have less number of parameters required to describe them.

Introduction:

In the Theory of Abstraction, we start from zero-postulation and build on to reach results that depend upon certain parameters that in turn depend on the system in question itself. Our theory seems to fit in reasonably well in our study of systems in all scales. Starting from a basis of no postulation, we build our theory. As we go on piling up possibilities, we come to a similar basis for understanding the four non-contact forces of nature known till date. The difference in ranges of these forces is explained from this basis in previous papers. Zero postulation or abstraction as the basis of theory synthesis allows us to explore even imaginary and chaotic non-favored solutions as possibilities. With no postulation as the fundamental basis, we are thus able to pile up

7

postulated results or favored results, but not the other way round. We keep describing such implications of abstraction in this paper too. We deal with the abstraction of observable parameters involved in the Standard Model.

From previous work on th theory, we know that the force that will be felt due to a quantity A' of the property A in this field of acceleration a is,

$$F =$$

$$A' c^2 \frac{(\Delta A)^{\frac{4i}{8i-1}}}{(A)^{\frac{2}{8i-1}}} \qquad \ldots (1)$$

where the acceleration a is,

$$a =$$

$$c^2 \frac{(\Delta A)^{\frac{4i}{8i-1}}}{(A)^{\frac{2}{8i-1}}} \qquad \qquad \ldots (2)$$

The uncertainty, in turn, depends upon the Lyapunov exponents (ν). Moreover, there is a stretching or shrinking of a given direction according to the factor $e^{\nu t}$, according as ν being positive or negative in that direction.

This will be our starting point in our study of the Standard Model in this present body of work.

<u>Higgs Potential:</u>

When the gauge coupling constants, depending upon equation (2), are g and g' for of SU(2) and U(1), respectively, the tree-level Higgs Potential in the minimal (SUSY) Standard Model is,

$$v = \frac{g^2}{8} (\bar{H}_1 \,_{\tau_a} H_1 + \bar{H}_2 \,_{\tau_a} H_2 \,)^2 + \frac{g'^2}{8} (\bar{H}_1 H_1 + \bar{H}_2 H_2 \,)^2$$

where (here),

$$\tau_a = \frac{A^{\frac{1}{8i-1}}}{c} \qquad \qquad ...(3)$$

From the breaking of electroweak symmetry caused by

Abstraction in Theory–Zero Postulation Results

$$< H_1 > = \begin{pmatrix} 0 \\ v_1 \end{pmatrix} / \sqrt{2}$$

and

$$< H_2 > = \begin{pmatrix} v_2 \\ 0 \end{pmatrix} / \sqrt{2}$$

we get the tree level masses of the Higgs scalars as,

$$m^2{}_{\chi^0} = m_1^2 + m_2^2,$$

$$m^2_{\chi^\pm} = m^2_{\chi^0} + m^2_{W^\pm},$$

$$m_{a,b}^2 = \frac{1}{2}[m_{x^0}^2 + m_{z^0}^2 \pm$$

$$\sqrt{\{(m_{x^0}^2 + m_{z^0}^2)^2 -}$$

$$4m_{x^0}^2 m_{z^0}^2 cos^2 2\theta\}]$$

$$\tan\theta = \frac{v_2}{v_1}.$$

Interacting Boson Models:

For a Hamiltonian $H(q, p)$ and equations of motion

$$\dot{q}_i = \frac{\partial H}{\partial p_i}, \dot{p}_i = \frac{\partial H}{\partial q_i}$$

With D degrees of freedom, $x = (q, p)$,

$$q = (q_1, q_2, q_3, \ldots, q_D),$$

$$p = (p_1, p_2, p_3, \ldots, p_D).$$

The value of the Hamiltonian function at the state space point $x = (q, p)$ is constant along the trajectory $x(t)$. Thus the energy along the trajectory $x(t)$ is constant,

$$
\begin{aligned}
\frac{d}{dt} H[q(t), p(t)] \\
&= \frac{\partial H}{\partial q_i} \dot{q}_i(t) + \frac{\partial H}{\partial p_i} \dot{p}_i(t) \\
&= \frac{\partial H}{\partial q_i} \frac{\partial H}{\partial p_i} - \frac{\partial H}{\partial p_i} \frac{\partial H}{\partial q_i} \\
&= 0
\end{aligned}
$$

The trajectories therefore lie on surfaces of constant energy or level sets of the Hamiltonian $[(q, p) : H(q, p) = E]$.

Given a smooth function $g(x)$, the standard map is,

$$x_{n+1} = x_n + y_{n+1}$$

$$y_{n+1} = y_n + g(x_n).$$

This is an area-preserving map. The corresponding n^{th} iterate Jacobian matrix is,

$$M^n(x_0, y_0) = \prod_{K=n}^{1} \begin{pmatrix} 1 + g'(x_K) & 1 \\ g'(x_K) & 1 \end{pmatrix} \quad \dots (4).$$

The complete Hamiltonian of the IBM1 is,

$$H = H^{(I)} + v_R R^2 + v_q Q^2,$$

where

$$H^{(I)} = \varepsilon_n N + v_n N^2 +$$
$$\left(\varepsilon_d' + v_{nd}N\right)n_d + v_d n_d^2 +$$
$$v_t T^2 + v_j J^2$$

$|N n_d \tau n_\Delta J M >$ are eigenfunctions of the operator $H^{(I)}$, with eigenvalues $E^{(I)}_{N\tau J n_d}$.

A Floquet multiplier $\Lambda = \Lambda(x_0, t)$ associated to a trajectory is an eigenvalue of the Jacobian matrix J and it satisfies

$$\det(J - \Lambda I) = \det(J^T - \Lambda I)$$
$$= \det(-\omega J^T \omega - \Lambda I)$$

$$= \det(J^{-1})\det(I - \Lambda J)$$

$$= \Lambda^{2D} \det(J$$
$$- \Lambda^{-1}I) \qquad\qquad ...(5).$$

This is because, $J^{-1} = -\omega J^T \omega$, J being symplectic. If Λ is an eigenvalue of J so are $\frac{1}{\Lambda}$, Λ^* and $\frac{1}{\Lambda^*}$. Real eigenvalues always come paired as $\Lambda, \frac{1}{\Lambda}$. The complex eigenvalues come in pairs $\Lambda, \Lambda^*, |\Lambda| = 1$, or in loxodromic quartets $\Lambda, \frac{1}{\Lambda}, \Lambda^*$ and $\frac{1}{\Lambda^*}$.

For a trajectory originating near $x_0 = x(0)$ with an initial infinitesimal displacement $\delta x(0)$, the flow transports the displacement $\delta x(t)$ along the trajectory $x(x_0, t) = f^t(x_0)$.

This infinitesimal displacement is transported along the trajectory $x(x_0, t)$, with time variation given by,

$$\frac{d}{dt}\delta x_i(x_0, t) =$$

$$\sum_j \frac{\partial v_i}{\partial x_j}(x)\bigg|_{x=x(x_0,t)} \delta x_j(x_0, t)$$

For two scalar bosons φ_a and φ_b the mass-matrix is,

$$(\varphi_a \varphi_b) \begin{bmatrix} m_{x^0}^2 + m_{z^0}^2 \sin^2 2\theta & -m_{z^0}^2 \sin 2\theta \cos 2\theta \\ +\frac{1}{2}\delta \sin^2 2\theta.v^2 & +\delta \sin^2 \theta \sin 2\theta.v^2 \\ -m_{z^0}^2 \sin 2\theta \cos 2\theta & m_{z^0}^2 \cos^2 2\theta \\ +\delta \sin^2 \theta \sin 2\theta.v^2 & +2\delta \sin^4 \theta.v^2 \end{bmatrix} \begin{pmatrix} \varphi_a \\ \varphi_b \end{pmatrix}$$

where, $\delta \simeq 3\left(log\frac{m^2+m_t^2}{m_t^2}\right)\left(\frac{h_t^2}{4\pi}\right)$.

One of the eigenvalues being always smaller

than $m_{z^0}^2 \cos^2 2\theta + 2\delta \sin^4 \theta . v^2$ the

mass of the lightest scalar boson φ_l is,

$$m_l \leq \sqrt{\{m_{z^0}^2 \cos^2 2\theta + \frac{6}{(2\pi)^2}\left(\log \frac{m^2+m_t^2}{m_t^2}\right)\frac{m_t^4}{v^2}\}}$$

The equations of motion for a time-independent D-degrees of freedom Hamiltonian can be written as,

$$\dot{x}_i = \omega_{ij} H_j(x), \omega$$
$$= \begin{pmatrix} 0 & -I \\ -I & 0 \end{pmatrix}, H_j(x)$$
$$= \frac{\partial}{\partial x_j} H(x);$$

where $x = (q, p) \in B$ is a phase space

point. $H_K = \partial_K H$ is the column vector of partial derivatives of H, I is the $[D \times D]$ unit matrix and ω the $[2D \times 2D]$ symplectic form.

$$\omega^\mathsf{T} = -\omega, \omega^2 = -1$$

The evolution of J^t is determined by the stability matrix A,

$$\frac{d}{dt} J^t(x) = A(x) J^t(x),$$

$$A_{ij}(x) = \omega_{ik} H_{kj}(x)$$

where the matrix of second derivations $H_{kn} = \partial_k \partial_n H$ is the Hessian matrix. For symmetry of H_{kn}, $A^T \omega + \omega A = 0$.

The eigenenergy of the I-th state is $E_{NJ}^{(I)}$ and

satisfies the relation,

$$H|INMJ \gg > = E_{NJ}^{(I)}|INMJ \gg .$$

The Yukawa coupling h_t of the top quark is

defined in the superpotential,

$$G = h_t \varphi_{tR} \varphi_{qL^3} \varphi_{H_2} ,$$

where φ_{tR} and φ_{qL^3} are chiral superfields

of the right-handed top quark and the left-

handed quark doublet in the third generation,

respectively.

In this respect, the Yukawa coupling of the top

quark is,

$$h_t = \sqrt{2}m_t/v_2 .$$

The Hamiltonian $H^{(1)}$ of an active boson of a single state $|b_{IM} >$ is constituted by its kinetic energy $K^{(1)}$ and its potential energy $P^{(1)}$.

The IBMs are seen to obey the Laws of Physical Transaction that follows from Zero-Postulation. The chaotic superfields at the requisite scaling-ratio yields necessary equation-parameters needed to describe them at that given scaling-ratio. This is seen to be independent of the choice of scale, but at smaller scaling-ratios, we have less loss of information. At a higher scale, we seem to have less number of parameters required to describe them.

The 12 creation operators for bosons are:

$$b^+_{\pi,jm} = s^+_\pi, d^+_{\pi,m} \ (m = -2, -1, \ldots, 2)$$

and

$$b^+_{v,jm} = s^+_v, d^+_{v,m} \ (m = -2, -1, \ldots, 2)$$

where π is a proton and v is a neutron.

Thus, for these pairs,

$$H = H_\pi + H_v + V_{\pi v} \ ,$$

which is the usual result.

A normalized, symmetric many-boson state (with occupational numbers: $N_a, N_b, \ldots, N_x)$, for creation operator b^+_n

(with index n), is represented by,

$$(N_a!\,N_b!\,...\,N_x!)^{-\frac{1}{2}}(b_a^+)^{N_a}(b_b^+)^{N_b}\,...\,($$
$$= |N_a N_b ... N_x >$$

where the index represents the angular momentum of the single state and its projection.

Conservation of Boson Number:

Let us now consider the flow of an energy quantum, with frequency (ν) in a given direction in vacuum. The distance (D) of transport in that given direction can be considered to be (cT); where (c) is the velocity of energy-quantum in vacuum and (T) is the time of transport.

For an empty environment in which the flow takes place, we may assume that the concerned difference in concentrations between the initial and the final points,

$$\lambda = h\nu$$

where h is Plank's constant.

Placing $D = cT$ and $\lambda = h\nu$ in the transaction equation, we get,

$$F \propto \frac{\omega h\nu S}{cR}$$

Again, as the energy-quantum moves from the initial to the final point completely, the concerned flow,

$$F = h\nu$$

Placing this value of F in the previous equation, we get,

$$\omega = c \left(\frac{R}{S} \right).$$

Assuming, the resistance against the concerned flow and the support towards it for the energy, i.e., considering $R = S$, we have,

$$\omega = c \qquad \ldots (8)$$

The value of the constant ω may therefore be replaced by the speed of light in vacuum, C, in the equations concerning transport of a given physical entity.

Such a transfer of any physical entity as

described by the transaction equation will continue until and unless the difference in concentrations concerned, i.e., λ becomes zero.

While

$$\frac{F}{T} = 2c\left(\frac{\lambda}{D}\right) \qquad \dots (9).$$

Equation (9) describes fundamentally the effect (i.e., the flow F in time T) of two material-points having same factorial conditions regarding one or a number of entities. Considering a collection of such points and applying a statistical approach, the logistic equation (due to May, 1967) for $\left(\frac{F}{T}\right)$ can be written as,

$$2c\left(\frac{\lambda}{D}\right)_{t+1} = 2Kc\left(\frac{\lambda}{D}\right)_t\left[1 - 2c\left(\frac{\lambda}{D}\right)_t\right]$$

i.e.,

$$\left(\frac{\lambda}{D}\right)_{t+1} = K\left(\frac{\lambda}{D}\right)_t\left[1 - 2c\left(\frac{\lambda}{D}\right)_t\right]$$

where K is a constant.

Also, the quadratic map (due to Lorentz, 1987) can be written as,

$$2c\left(\frac{\lambda}{D}\right)_{t+1} = K - \left(2c\frac{\lambda}{D}\right)_t^2$$

i.e.,

$$2c\left(\frac{\lambda}{D}\right)_{t+1} = K - 4c^2\left(\frac{\lambda}{D}\right)_t^2 \quad \ldots (10).$$

Let us consider a system of n_d bosons. The parameter d represents its structural-orientation, which depends upon our experimental conditions (like the scaling-ratio used, differences in concentration, etc.). For our present scale of interest, the frequency of a boson ϑ_d represents the given difference in concentration λ_d .

Thus the total energy carried by a given boson is,

$$E = h\lambda_d$$

Also, let us have n_{d_1} and n_{d_2} bosons interacting, such that total energy in the system is,

$$E = h\left(\lambda_{d_1} + \lambda_{d_2}\right).$$

Abstraction in Theory–Zero Postulation Results

The principle of conservation of energy for that given system requires that the resulting system of bosons (λ_{d_3}) have an equal amount of total energy, such that,

$$\lambda_{d_3} = \frac{\lambda_{d_1} + \lambda_{d_2}}{2},$$

which, in turn, means,

$$n_{d_3} = n_{d_1} + n_{d_2}.$$

All trajectories described by the quadratic map become asymptotic to $-\infty$ for $K < -0.25$ and $K > 2$.

In the SO(6) or γ-instable limit, a linear combination of Casimir operators on the chain

$$U(6) \supset SO(6) \supset SO(5) \supset SO(3)$$

is,

$$H = \varepsilon_n N + v_n N^2 + v_r R^2 + v_t T^2 + v_j J^2$$

For the functions $\varphi_i = \lambda_i$ and $\varphi_j = \lambda_j$ belonging to the same system,

$$\int \lambda_j^* (\underline{R}) \lambda_i(\underline{R}) d\underline{R}$$

$$\equiv \iint \lambda_j^* (\underline{R}) \delta(\underline{R} - \underline{R}') \lambda_i(\underline{R}') d\underline{R} d\underline{R}'$$

$$= \sum_\beta \int \lambda_j^* (\underline{R}) \lambda_\beta(\underline{R}) d\underline{R}$$

$$\cdot \int \lambda_\beta^* (\underline{R}') \lambda_i(\underline{R}') d\underline{R}' \qquad \dots (11).$$

Using Lyapunov exponents for a given transport, and replacing $2c\left(\dfrac{\lambda}{D}\right)$ by a quantity $'\tau'$, we have,

$$\frac{d}{d\tau} f^n(\tau) = \frac{\delta n}{\delta o}$$

i.e.,

$$\frac{\delta n}{\delta o}$$

$$= \prod_{i=1}^{n} f'(\tau_i) \qquad \dots (12)$$

$$b = \frac{1}{n} \log_e \left(\frac{\delta n}{\delta o} \right)$$

i.e.,

$$b$$

$$= \frac{1}{n} \sum_{i=1}^{n-1} \log_e \left| f'(\tau_i) \right| \qquad \dots (13),$$

where b is a constant (the local slope of all possible routes), and

$$\Psi = \lim_{n \to \infty} \frac{1}{n} \sum_{i=0}^{n-1} \log_e \left| f'(\tau_i) \right| \quad \dots (14),$$

where Ψ is a constant for the system.

Non-commutative Operators:

Let the spectral triple $(\mathcal{A}, \mathcal{H}, \mathcal{D})$ denote a non-commutative geometry, where in Hilbert space $\mathcal{H}$ we have an involutive geometry $\mathcal{A}$ and a self-adjoint unbounded operator $\mathcal{D}$. $\mathcal{D}^{-1}$, the inverse of $\mathcal{D}$ denotes infinitesimal length. $\mathcal{A} = C^{\infty}(M)$ is the algebra of smooth functions on the Riemann Manifold M. The Dirac operator of Levi-Civita spin is $\mathcal{D}$.

The Hilbert space of L^2-spinors is,

$$\mathcal{H} = L^2(M, S).$$

Considering x dimensions of the manifold M, a $Z/2$ gradient for difference in concentrations of one or a group of physical quantity or quantities λ_x in the Hilbert space $\mathcal{H}$ satisfies:

$$\lambda_x = \lambda_x^*, \lambda_x^2 = 1, \lambda_x a = a\lambda_x \; \forall \; a \in \mathcal{A}, \lambda_x \mathcal{D} = -\mathcal{D}\lambda_x \; ... \; (15)$$

.

The spectrum of the operator $\mathcal{D}$ replaces the status of points $x \in M$ in commutative geometry.

x modulo 8 determines the values of the parameters $\varepsilon, \varepsilon', \varepsilon''$ such that,

$$J^2 = \varepsilon, JD = \varepsilon' DJ, J\lambda_x = \varepsilon'' \lambda_x J,$$
$$\varepsilon, \varepsilon', \varepsilon'' \in \{-1,1\}$$

,

where J is an anti-linear isometry in the Hilbert space $\mathcal{H}$ and it denotes the real structure on $\mathcal{H}$.

For any given flow F in $\mathcal{H}$, we have,

$$\mathcal{A} = C^\infty(M) \otimes \mathcal{A}_F ,$$

$$\mathcal{A}_F = C \oplus \mathcal{H} \oplus M_3(C),$$

$$\mathcal{H} = \{\begin{pmatrix} \alpha & \beta \\ -\overline{\beta} & \overline{\alpha} \end{pmatrix}; \alpha, \beta \in C\},$$

$$\mathcal{H} = L^2(M, S) \otimes \mathcal{H}_F, \mathcal{D} = \eth_M \otimes 1 + \lambda_{x5} \otimes \mathcal{D}_F$$

The Dirac operator $\mathcal{D}_x$, considering inner fluctuations, yields a 36×36 matrix for the 36 quarks. σ^α and ψ^i represent Pauli matrices and Gell-Mann matrices, respectively. This matrix, with Clifford algebra tensors, is,

$$\mathcal{D}_x =$$

$$
\begin{bmatrix}
\lambda_x^\mu \otimes \left(\mathcal{D}_\mu 1_2 - \frac{i}{2} g_{02} A_\mu^\alpha \sigma^\alpha - \frac{i}{6} g_{01} B_\mu \otimes 1_2 \right) \otimes 1_3, & \lambda_{x5} \otimes K_0^d \otimes \mathcal{H}_F, & \lambda_{x5} \otimes K_0^u \otimes \widetilde{\mathcal{H}}_F \\[2ex]
\begin{aligned} \lambda_{x5} \otimes K_0^{d^*} \otimes \mathcal{H}_F^*, \\ \lambda_{x5} K_0^{u^*} \widetilde{\mathcal{H}}_F^*, \end{aligned} & \begin{aligned} \lambda_{x5} \otimes \left(\mathcal{D}_\mu + \frac{i}{3} g_{01} B_\mu \right) \otimes 1_3, \\ 0, \end{aligned} & \begin{aligned} 0 \\ \lambda_{x5}{}^\mu \otimes \left(\mathcal{D}_\mu - \frac{2i}{3} g_{01} B_\mu \right) \otimes 1_3 \end{aligned}
\end{bmatrix}
$$

$$\otimes 1_3 + \lambda_{x5}{}^\mu \otimes 1_4 \otimes 1_3 \otimes \left(-\frac{i}{2} g_{03} V_\mu^i \psi^i \right),$$

where, B_μ , A_μ^α and V_μ^i are the $U(1)$, $SU(2)_w$ and $SU(3)_c$ gauge fields, respectively, with gauge couplings g_{01}, g_{02} and g_{03}. $\widetilde{H_F} = i\sigma^2 H_F$.

A similar treatment, after considering inner fluctuations, yields a 9×9 matrix for leptons,

$$D_\chi$$

$$= \begin{bmatrix} \lambda_\chi^\mu \otimes \left(\mathcal{D}_\mu - \frac{i}{2} g_{02} A_\mu^a \sigma^a - \frac{i}{2} g_{01} B_\mu \otimes 1_2 \right) \otimes 1_3 & \lambda_{\chi 5} \otimes K_0^e \otimes \mathcal{H}_F \\ \lambda_{\chi 5} \otimes K_0^{e*} \otimes \mathcal{H}_F^* & \lambda_{\chi 5}^\mu \otimes \left(\mathcal{D}_\mu + i g_{01} B_\mu \right) \otimes 1_3 \end{bmatrix}$$

Conclusion:

We study the Standard Model taking into consideration the Zero-Postulation of the Theory of Abstraction. The relative difference in concentrations of any given physical entity creates a tensor-gradient that causes the varied ways of flow. The necessary complete set of parameters and the required scaling-ratio for a given set of observations describes the observations themselves. We can arrive at

IBMs and IBFMs in this way. We may as well describe the basis of the Standard Model itself using the Theory of Abstraction.

2 Abstraction and Structures in Energy

Key Ideas:

Zero postulation and the principles of the Theory of Abstraction are used to study structures of energy inside a black hole, which is incredibly heavy and incredibly small. We chase the questions, how matter (with various structures) is formed from energy and the energy making up matter has to be in what orientation to form the matter that we see. We arrive at the fundamental model and the equations describing the formation of structure in energy.

Introduction:

The Theory of Abstraction and the principle of Zero Postulation describes everything that there 'is' in the universe (including spacetime) as an expression of energy. It explains directivity towards optimized solutions and cluster formation in the universe as inherent properties of the energy itself. It must also be able to describe the formation of various structures in energy. As such, it is of great interest for us to investigate how the matter that we see around us is formed from the various orientations of the structured formed in energy.

A theory that is able to describe the world in totality has to keep the number of basic postulates it depends upon to zero or near zero. Reductionism hits a dead end in this regard. On the other hand, abstraction as the starting point of building up a theory may be seen to be of fitting use. It would be much more than a new way of tackling the problem. Even abstract postulates do away with the shackles that bind our theories into the system and bar them from being total descriptions of the system. The abstraction we are talking about here may be defined as, "Postulation of non-postulation" or, in other words, "A system of postulation that gives equal weights to all possible solutions inside the system and favors none of such solutions over others."

Abstraction automatically gives rise to optimized solutions within the universal set of all possible solutions, as has been shown in this book. It is these optimized solutions that make up and drive the non-abstract parts of the world, while the non-optimized solutions remain 'hidden' from the material world, inside the abstract world.

Starting from a basis of no postulation, we build our theory. As we go on piling up possibilities, we come to a similar basis for understanding the four non-contact forces of nature known till date. The difference in ranges of these forces is explained from this basis in this book. Zero postulation or abstraction as the basis of theory synthesis allows us to explore even imaginary and chaotic non-favored solutions as possibilities. With no postulation as the fundamental basis, we are thus able to pile up postulated results or favored results, but not the other way round. We keep describing such implications of abstraction in this book. We deal with the abstraction of observable parameters involved in a given system (quantum, relativistic, chaotic, non-chaotic) and formulate a similar basis of understanding them.

Let us consider the example of a three-point isolated system. Let the points be 'A', 'B' and 'C'. Let A and B be material points, whereas, C be situated anywhere on the straight line joining A and B. The material parts of both A and B tends to move in all possible directions. These possible directions include the directions towards each other. Thus, at point C, for obvious reasons, an additional effect will be

43

felt due to the tendency of material to flow from A to B and from B to A, as compared to all other directions.

The points A, B and C being considered parts of an isolated system and all three points being assumed fundamentally similar (with the only difference that A and B contain material, while C is empty), the factors R and S must be equal.

Thus, we have:

$$\frac{F}{T} = c\frac{\lambda}{D}$$

D being considered the 'distance' between A and C and x the distance between A and B (say), the distance between B and C is $D - x$. This distance can be any length of any given dimensions (as determined by the scaling-ratio of observations) between two points in spacetime.

The effect on C due to the material-point A can thus be written as,

$$\frac{F_A}{T_A} = c\,\frac{\lambda_A}{x}$$

Similarly, the effect on the empty point C due to the material-point B is,

$$\frac{F_B}{T_B} = c\,\frac{\lambda_B}{D - x};$$

where F_A and F_B are the respective values of flows towards the point C due to A and B, respectively. T_A and T_B are the respective values of time and λ_A and λ_B are the respective values of the differences in concentrations of the concerned entity between A and B.

Substituting x in the above two equations, we have,

$$c\,\frac{T_A\lambda_A}{F_A} = D$$

$$-\,c\,\frac{T_B\lambda_B}{F_B} \qquad\qquad ...\,(1)$$

Considering the points to be having equal factors, i.e., considering $\lambda_A = \lambda_B = \lambda$ (say), $F_A = F_B = F$ (say) and $T_A = T_B = T$ (say), equation (1) reduces to,

$$c\,\frac{T\lambda}{F} = D - c\,\frac{T\lambda}{F}$$

i.e.,

$$\frac{F}{T} = 2c\left(\frac{\lambda}{D}\right) \qquad \ldots (2)$$

Equation (2) describes fundamentally the effect (i.e., the flow F in time T) of two material-points having same factorial conditions regarding one or a number of entities. Considering a collection of such points and applying a statistical approach, the logistic equation for $\left(\frac{F}{T}\right)$ can be written as,

$$2c\left(\frac{\lambda}{D}\right)_{t+1} = 2Kc\left(\frac{\lambda}{D}\right)_{t}\left[1 - 2c\left(\frac{\lambda}{D}\right)_{t}\right]$$

i.e.,

$$\left(\frac{\lambda}{D}\right)_{t+1} = K\left(\frac{\lambda}{D}\right)_{t}\left[1 - 2c\left(\frac{\lambda}{D}\right)_{t}\right] \qquad \ldots (3)$$

where K is a constant.

Also, the quadratic map can be written as,

$$2c \left(\frac{\lambda}{D}\right)_{t+1} = K - \left(2c\frac{\lambda}{D}\right)_t^2$$

i.e.,

$$2c \left(\frac{\lambda}{D}\right)_{t+1} = K$$
$$- 4c^2 \left(\frac{\lambda}{D}\right)_t^2 \qquad \ldots (4)$$

All trajectories described by the quadratic map become asymptotic to $-\infty$ for $K < -0.25$ and $K > 2$

As we deal with the flow of a given material entity towards one given point or the effects on a given point, the expression for the attractor for each such point can be written as,

$$\left(2c\frac{\lambda}{D}\right)^* = \left(1 - \frac{1}{K}\right) \qquad ...(5);$$

where $0 < K < B$.

$\left(2c\frac{\lambda}{D}\right)^*$ is a point in the desired

dimensional plot into which the trajectories seem to crowd. As we do not need to deal with more than one attractor or periodic point, the trajectories will tend to revisit only the attractor point concerned, to the desired level of accuracy of observations and calculations.

For $K \geq 3$, the trajectory behaviour becomes increasingly sensitive to the value of K. There are a few more points to be noted regarding the dependence of the trajectory behavior on the values of K:

1. For $K \leq 1$, the attractor is a fixed point and has a value 0.

2. For $1 < K < 3$, the attractor is a fixed point and its value is > 0 but < 0.667.

3. For $3 \leq K \leq 3.57$, period doubling occurs, with the attractor consisting of $2, 4, 8$, etc., periodic points as K increases within that range.

4. For $3.57 < K \leq 4$, we have the region of chaos, where the attractor can be erratic (chaotic with infinitely many points) or stable.

For all calculations, the desired conditions may

be placed at the attractor. A trajectory never gets completely and exactly all the way into an attractor though, but only approaches it asymptotically. In the region of chaos, we apply the method of searching for windows or zones of K-values for which iterations from any initial conditions will produce the periodic attractor, instead of a chaotic one. For the logistic equation(3), the most common such zone lies at $K \approx 3.83$ and for the quadratic map(4), at $K \approx 1.76$.

Inside A Black Hole:

A black hole may be considered to be a system, whose packing-density has a certain minimum value. If the black hole's mass is M, which is concentrated within a radius R then the minimum value of the ratio

$$\frac{M}{R} = \frac{c^2}{2G} \approx 10^{27} kg/m$$

Be it the large vastness of the universe or the delicate smallness of the sub-atomic world, by choosing a suitable constant scaling ratio for both, we may obtain their representations. These representations, following a certain constant scaling ratio, will be self-same. In the previous chapters, I have mentioned the chaotic behavior in the quantum world. Choosing suitable scaling ratios, we may turn the universe itself into such a chaotic quantum system, having its own necessary quantum states and trajectory behavior. In that case, the study of the universe reduces to the study of some sort of a quantum chaotic system.

On the other hand, choosing some other necessary scaling ratios, the atomic and the sub-atomic realm may be extended to become the universe itself, complete with its own macroscopic trajectory behavior. Instead of formulating different ways of looking at worlds of different sizes, if we adjust the way of viewing i.e., the scaling ratios in such a fashion that the representations of the worlds merge, we will be looking at representative worlds of study which are practically self-same. The Laws of Physical Transactions formulated in previous chapters may then be applied in order to study such self-same representations of the worlds of various scales. Unification of the ways of studying at different ranges of scaling may thus be achieved by suitable landscaping (adjusting different scales to a suitable scaling ratio, in order to make all the scales of study similar in size).

Further, a similar approach may be applied to study the Bose-Einstein Condensation. A certain critical packing density of the constituents of each world of a certain landscape must ensure a condensation of similar sort. The quantum states (or some similar states) of each such landscape will merge and give spikes for that critical scaling ratio in their respective representations.

The Abstraction Theory is applied in landscaping. A collection of objects may be made to be vast or meager depending upon the scale of observations. This idea may be developed to unite the worlds of the great vastness of the universe and the minuteness of the sub-atomic realm. Keeping constant a scaling ratio for both worlds, these may actually be converted into two self-same representatives with respect to scaling. The Laws of Physical Transactions are made use of to study Bose-Einstein condensation. As the packing density of concerned constituents increase to a certain critical value, there may be evolution of energy from the system.

Looking at a large enough part of the universe, we may draw an analogy to a system of scattered particles, in motion or rest, relative to each other. These particles may or may not be similar to each other, if we look at a given locality. Our idea, however, is that we can always represent even the whole of the universe on a piece of paper of our desired size. We can very well do the same with localities of sub-atomic sizes.

We may represent both the worlds, viz., the

microscopic and the macroscopic, within any desired standard size. Theoretically, we are only to diminish the snaps of the universe and magnify the snaps of the microscopic world in order to put both into representations of a definite scaling-size. Looking at such a representation of the macroscopic world (due to the large number of constituents and the large distances separating them involved) we will find it to be a complex mixture of various kinds of particles. On the other hand, looking at such a representation of the microscopic world, (due to the small distances separating the constituents) it will be like the actual universe itself, with various types of constituent parts involved. Such a representation of the microscopic and the macroscopic worlds will bring out hidden properties and behaviors of both worlds, as well as providing for a similar basis of studying them both.

Let us consider a given representation with fractal dimension D_F. The fractal dimension is purely geometrical, i.e., it only depends on the shape of the representation. A suitable probability measure $d\mu$, according to the particular phenomenon considered is assigned to the given representation. A coarse grained probability density, as the mass of the hypercube Λ_i of size l is defined as,

$$P_i(l) = \int_{\Lambda_i} d\mu(x)$$

where $i = 1, 2, 3, \ldots, N(l)$.

The information dimension D_I is such that,

$$\sum_{i=1}^{N(l)} P_i \ln(P_i) \simeq D_I \ln(l) \qquad \dots (6);$$

where $D_I \leq D_F$.

The number of boxes containing the dominant contributions to the total mass and thus relevant part of the information, is,

$$N_R(l) \propto l^{-D_I} \qquad \dots (7).$$

For each box Λ_i, $D_I = D_F$ for a uniform distribution. When $D_I < D_F$, the measure itself may be called fractal since it is singular with respect to the uniform distribution,

$$P^* = \frac{1}{N(l)} \propto l^{D_F}$$

For each box Λ_i. Thus, $\dfrac{P_i}{P_i^*}$ can diverge in the limit of vanishing l.

Simulations of the mass-moment scaling yields,

$$\langle P_i(l)^q \rangle \equiv \sum_{i=1}^{N(l)} P_i(l)^{q+1}$$
$$\propto l^{q \cdot d_{q+1}} \qquad \ldots (8).$$

The d_q are the Renyi dimensions which generalize the information dimension $D_I = d_1$ as well as the fractal dimension $D_F = d_0$. If the d_q's are not constant, anomalous scaling is to be employed and, as the order q varies, the amount of the difference $D_q - D_F$ gives a first rough measure of the heterogeneity of the probability distribution.

The moment generic observables A computed on scale l is such that,

$$\langle A(l)^q \rangle \propto l^{g(q)} \qquad \ldots (9)$$

Anomalous scaling, i.e., a non-linear shape of the function $g(q)$ is the more common situation, where one does not require unnecessarily to consider only a finite number of scaling components. In some cases, one may observe strong time variations in the degree of chaoticity. This intermittency phenomenon involves an anomalous scaling with respect to time-dilations identifying the parameter e^{-t} with the parameter l used in spatial dialations. A measure of the degree of intermittency requires the introduction of infinite sets of exponents which are analogous to the Renyi dimensions and can be related to a multifractal structure given by the dynamical system in the functional trajectory space.

The Grassberger-Procaccia correlation dimension ν is defined by considering the scaling of the correlation integral,

$$C(l) = \lim_{M \to \infty} \frac{1}{M^2} \sum_{i} \sum_{j \neq i} \theta \, (l - |x_i - x_j|);$$

where θ is the Heaviside step function and $C(l)$ is the percentage of pairs (x_i, x_j) with distance $|x_i - x_j| \leq l$.

In the limit $l \to 0$, $C(l) \propto l^\nu$.

In   general,

$$v \leq D_F.$$

v is a more relevant scaling index than D_F since it is related to the point probability distribution on the attractor, while D_F cannot take into account an eventual homogeneity in the visit frequencies.

Let us define the number of points in an F-dimensional spherical representation of the world, with radius l and centre at x_i as,

$$n_i(l) = \lim_{M \to \infty} \frac{1}{M-1} \sum_{j \neq i} \theta \, (l - |x_i - x_j|) \qquad \ldots (10).$$

We must introduce a whole set of generalized scaling exponents,

$$\langle n(l)^q \rangle$$

$$= \lim_{M \to \infty} \frac{1}{M} \sum_{i=1}^{M} n_i(l)^q$$

$$\propto l^{\emptyset(q)}$$

where $\emptyset(1) = \nu$.

Considering a uniform partition of phase space into boxes of size l it is convenient to introduce the probability $P_K(l)$ that a point x_i falls into the Kth box. In this case, the moments of P_K can be estimated by summing up the boxes,

$$\langle p(l)^q \rangle = \sum_{K=1}^{N(l)} P_K(l)^{q+1}$$

$$\propto l^{q.d_{q+1}} \qquad \ldots (11)$$

A moment of reflection shows,

$$\emptyset(q)/q = d_{q+1}$$

because of the ergodicity $n_i(l) \sim P_K(l)$, if x_i belongs to the Kth box and since one can use either an 'ensemble' average (weighted sum over the boxes) or a 'temporal' average (sum of the time evolution $x(l)$).

The fractal dimension for $q = -1$ is,

$$D_F = d_0$$
$$= -\emptyset(-1)$$

while the correlation dimension is,

$$v = d_2$$
$$= \emptyset(1)$$

According to the Theory of Physical Abstraction, each point x should have the same singularity structure,

$$\Delta V_x(r) \propto r^h, h$$
$$= \frac{1}{3} \qquad \ldots (12)$$

In other words $\varepsilon(x)$ tends to be smoothly distributed in a region of R^3. The eddy turn-over time and the kinetic energy per unit mass at scale r are defined as,

$$t(r) \sim \frac{r}{\Delta V(r)} \qquad \ldots (13)$$

and

$$E(r) \sim \Delta V(r)^2 \qquad \ldots (14)$$

The transfer rate of energy per unit mass from the eddy at scale r to smaller eddies is then given by

$$\tilde{\varepsilon}(r)$$

$$= \frac{E(r)}{t(r)} \sim \frac{\Delta V(r)^3}{r} \qquad \ldots (15)$$

Since

$$\varepsilon_x(r) = \left(\frac{1}{r^3}\right) \int_{\Lambda_x(r)} \varepsilon(y) d^3 y,$$

$[\Lambda_x(r)$ is a cube of edge r around x we have,

$$\int_{\Lambda_x(r)} \varepsilon(y) d^3 y \sim r^3 \qquad \ldots (16)$$

$r \to 0$ means r in the initial range and the regions containing a large part of $\varepsilon(x)$ are a physical approximation of a fractal structure. In this β —model approach,

$$\int_{\Lambda_x(r)} \varepsilon(y)\,d^3y \propto \begin{cases} r^{D_F} & if\ x \in S \\ 0 & if\ x \notin S \end{cases}$$

in an equivalent way,

$$\Delta V_x(r) \propto \begin{cases} r^h & if\ x \in S \\ 0 & if\ x \notin S; \end{cases}$$

where $h = (D_F - 2)/3$

At scale r, there is only a fraction,

$$r^{3-D_F} \propto \frac{r^{-D_F}}{r^{-3}}$$

occupied by active eddies.

The transfer energy from the eddy at scale l_n

(active eddy) to the scale l_{n+1} is $\varepsilon_n \propto \dfrac{v_n^3}{l_n}$.

Since, the energy transfer rate is constant in the

cascade process, for $\beta = 2^{D_F - 3}$, we have,

$$\varepsilon_n = \beta \varepsilon_{n+1}, \frac{v_n^3}{l_n}$$

$$= \beta \frac{v_{n+1}^3}{l_{n+1}} \qquad \dots (17)$$

Iterating, we have,

$$v_n$$

$$\propto l_n^{1/3} (l_n / l_0)^{\frac{D_F - 3}{3}}$$

Each eddy at scale l_n is divided into eddies of scale l_{n+1} in such a way that the energy transfer for a fraction β of eddies increases by a factor $\frac{1}{\beta}$, while it becomes zero for the other ones.

In order to generalize the β-model, we have at scale l_n, N_n active eddies. Each eddy $l_n(k)$ generates active eddies covering a fraction of volume $\beta_{n+1}(k)$. k labels the mother-eddy and $k = 1, ..., N_n$.

Since the rate of energy transfer is constant among mother-eddies and their effects, we have,

$$\frac{\nu_n(k)^3}{l_n}$$

$$= \beta_{n+1}(k) \frac{\nu_{n+1}(k)^3}{l_{n+1}} \qquad ... (18)$$

The iteration of ν_n gives an eddy generated by a particular history of fragmentations $[\beta_1, ..., \beta_n]$, such that,

$$\nu_n$$

$$\propto l_n^{1/3} \left(\prod_{i=1}^{n} \beta_i \right)^{-1/3} \qquad \ldots (19)$$

The fraction of volume occupied by an eddy generated by $[\beta_1, \ldots, \beta_n]$ is $\prod_{i=1}^{n} \beta_i$, such that,

$$\langle |\Delta V(l_n)|^P \rangle$$

$$\propto l_n^{P/3} \int \prod_{i=1}^{n} d\beta_i \ \beta_i^{(1-P/3)} P(\beta_1, \ldots, \beta_n)$$

With no correlation among different steps of the fragmentation, i.e., with $P(\beta_1, \ldots, \beta_n) = \prod_{i=1}^{n} P(\beta_i)$, the exponent concerned,

$$\zeta_P$$

$$= \frac{P}{3}$$

$$- \ln_2\left\{\beta^{(1-P/3)}\right\} \qquad \ldots (20)$$

Let us now consider a given representation of the universe. Let the packing density of the constituents be $\emptyset$. This packing density function $\emptyset$ will affect any given constituent point inside it in accordance with the Laws of Physical Transactions. The given constituent point concerned will in turn affect $\emptyset$ while interacting. For a given critical state of study of the total effects, we therefore are going to have a shear stress $\emptyset$ and a mean effective stress f. The critical state line is the loci of critical state conditions in the $\varepsilon - f - \emptyset$ space. Its projection on the $f - \emptyset$ space defines a strength parameter,

$$M = \frac{\emptyset}{f}$$
$$= \frac{6 \sin \emptyset}{3 - \sin \emptyset} \qquad \ldots (21).$$

The second equality applies to axis-symmetric, axial compression and it is a function of the constant volume critical state packing density function $\emptyset$.

The small-strain stiffness of a given representation is measured by imposing a smaller strain than the elastic threshold strain concerned. In this range, deformations localize at inter-point contacts and the granular skeleton deforms at constant fabric of spacetime. The nonlinear load-deformation response determines the stress-dependent shear wave velocity,

$$V_S \propto \left(\frac{f - \emptyset}{\varepsilon} \right)^{\beta} \qquad \ldots (22)$$

Inside a given cluster, we may have various growth-patterns. The growth may occur mainly at an active zone on the surface of the cluster. For a one-dimensional interface, a fluctuation-dissipation theorem exists, leading to an exact dynamic exponent $Z = \frac{3}{2}$. This is in excellent agreement with numerical simulations of ballistic aggregation and Eden clusters. For two-dimensional interfaces, $z \sim 1.5$.

The interface profile is described by a height $h(x, t)$. The simplest nonlinear Langevin equation for a local growth of the profile is,

$$\frac{\partial h}{\partial t}$$

$$= m\Delta^2 h + \frac{\lambda}{2}(\nabla h)^2$$

$$+ \eta(x, t) \qquad \dots (23)$$

The first term on the right-hand side describes relaxation of the interface by a tension term m. The second term is the lowest -order nonlinear term that can appear in the interface growth equation. Higher-order terms may also be present, but they are irrelevant and will not modify the scaling properties concerned. The noise $\eta(x, t)$ has a Gaussian distribution with $\langle \eta(x, t) \rangle = 0$ and

$$\langle \eta(x, t)\eta(x', t') \rangle = 2D\delta^d(x - x')\delta(t - t').$$

There is also a velocity term, but it is removed by choice of an appropriate moving coordinate system. Equation (23) is invariant under translation $h \to h + \text{constant},$ and obeys the infinitesimal reparametrization,

$$h \to h + b.x, x \to x + \lambda bt,$$

which describes the tilting of the interface by a small angle. When a given constituent point is added, the increment projected along the h-axis is,

$$\delta h = m\sqrt{[1 + (\nabla h)^2]}$$
$$\simeq m + (m/2)(\nabla h)^2 + \cdots$$

Following the transformation $W(x, t) = e^{[(\lambda/2m)h(x,t)]}$, we have,

$$\frac{\partial W}{\partial t}$$
$$= m\nabla^2$$
$$+ \left(\frac{\lambda}{2m}\right)\eta(x,t)W \qquad \ldots (24)$$

which is a diffusion equation in a time-dependent random potential. $W(x,t)$ is the sum of Boltzmann weights for all static configurations of a flow in a $(d+1)$-dimensional space from $(0,0)$ to (x,t). The noise term describes a quenched random potential $(\lambda/2m)\eta(x,t)$ exerted by the environment. The second transformation, $m = -\nabla h$ results in

$$\frac{\partial m}{\partial t} + \lambda m . \nabla$$
$$= m\nabla^2 m$$
$$- \nabla\eta(x,t)$$

which is the Burger's equation for a vorticity-free velocity field for $\lambda = 1$. In the Burger's equation, further evolution of the pattern proceeds through the larger parabolas growing at the expense of the smaller ones, and parallels the evolution of shock waves.

If the initial profile is $h(x, 0) = h_0(x)$, its evolution is given by,

$$
\begin{aligned}
& h(x, t) \\
& = \frac{2m}{\lambda} \ln \left\{ \int_{-\infty}^{\infty} \frac{d^d \xi}{(4\pi mt)^{\frac{d}{2}}} e^{\left[-\frac{(x-\xi)^2}{2mt} + \frac{\lambda}{2m} h_0(\xi) \right]} \right\}
\end{aligned}
$$

Let a given representation have bonds within itself, occupied by a resistance generated inside it due to its packing density $\emptyset$, with probability p. Let it have a support towards the concerned flow with probability $1 - p$. In such a representation, we have,

$$\left\langle \sum_k \varepsilon_k^n \right\rangle_{\xi, L} \propto L^{-x_n}, L$$

$$\rightarrow \infty \qquad \dots (25);$$

where $\left\langle \sum_k \varepsilon_k^n \right\rangle$ refers to the average over the sample realizations, L is the system-size $L \lesssim \xi$ and $\xi \propto (p - p_c)$ is the correlation length. ε_k is the energy dissipated in the branch k.

For a finite size scaling behaviour,

$$p\left(\sum_k \varepsilon_k^0, \sum_k \varepsilon_k^1, \dots, \xi, L \right)$$

$$= \lambda^{x_0} \lambda^{x_1} \dots p\left(\sum_k \varepsilon_k^0 / \lambda^{-x_0}, \frac{\sum_k \varepsilon_k^1}{\lambda^{-x_1}}, \dots, \xi/\lambda, \frac{L}{\lambda} \right) \dots (26),$$

(λ is the rescaling parameter) equation (25) implies,

$$\left\langle \left(\sum_k \varepsilon_k^n \right)^m \right\rangle_{\xi,L} \propto L^{-m x_n} \qquad \dots (27)$$

In disordered representations, the fluctuations of the free energy among different replicas may be regarded as the analogue of the temporal intermittency in a chaotic signal. Considering a spin-model of the D-dimensions, the Hamiltonian,

$$H\big[\{J_{ij}\}\big] = -\sum_{(i,j)} J_{ij}\sigma_i\sigma_j,$$

where $\sigma_i = \pm 1$ is the of the spin on the site

i and the coupling J_{ij} is an independent random variable distributed according to a probability distribution $p(J_{ij})$. Given a coupling realization $\{J_{ij}\}$, the partition function of an N spin system is the trace of the Boltzmann weight $e^{(-\beta H_N)}$,

$$Z_N(\beta, \{J_{ij}\}) = \sum_{\{\sigma_i\}} e^{\{-\beta H_N[\{J_{ij}\}]\}}$$

The free energy per spin in the limit $N \to \infty$ is,

$$F(\beta) = \lim_{N \to \infty} -\frac{1}{N\beta} \langle \ln Z_n \rangle$$

$$= \lim_{N \to \infty} -\frac{1}{N\beta} \int p(J_{ij}) dJ_{ij} \ln Z_N(\beta, \{J_{ij}\})$$

The free energy per spin of a coupling

realization $\{J_{ij}\}$ of a N spin system is,

$$\Xi_N$$

$$= -\frac{1}{N\beta} \ln Z_N\left(\beta, \{J_{ij}\}\right)$$

The self-averaging of Ξ_N is

$$F = \lim_{N\to\infty} \Xi_N$$

For a unidimensional system with first neighbor interactions and uniform field h, we can write the partition function as the trace over 2×2 random transfer matrices product. The Hamiltonian is now $H = -\sum_i (J_i \sigma_i \sigma_{i+1} + h\sigma_i)$, such that,

$$Z_N = Tr \prod_{i=1}^{N} M_i, M_i$$

$$= \begin{pmatrix} e^{\beta J_i + \beta h} & e^{-\beta J_i + \beta h} \\ e^{-\beta J_i - \beta h} & e^{\beta J_i - \beta h} \end{pmatrix} \qquad \ldots (29)$$

The moments of the partition function can be estimated as an integral over the spectrum of the possible free energies $[\Xi_{min}, \Xi_{max}]$,

$$\langle Z_N(\beta)^q \rangle \propto \int \prod (\Xi) \, d\Xi \; e^{(-\beta \Xi q N)} \qquad \ldots(30)$$

The Kolmogorov entropy is related to the sum of the positive Lyapunov exponents which measure the divergence rate along the expanding directions, in accordance with the Theory of Physical Abstraction. For an ergodic measure with a compact support (as proved by Pesin) is,

$$K_1 \leq \sum_{i=1}^{P} \lambda_i \; ;$$

where P is the number of exponents,

$\lambda_i > 0$. In Hamiltonian systems,

$$K_1 = \sum_{i=1}^{P} \lambda_1 = \left. \frac{dL^{(P)}}{dq} \right|_{q=0}$$

A record of measures of a signal $x(t)$ at uniform spacing τ is

$$x_i = x(i\tau); i = 1, 2, \ldots, M$$
$$\gg 1 \qquad \ldots (31)$$

Clustering:

Since the number of eddies at scale l with singularity h is proportional to $l^{-d(h)}$, the number of grid points that have to be considered for resolving the set $S(h)$ is

$$N_h(R_e) \sim (L/\eta(h))^{d(h)}$$
$$\propto R_e^{d(h)/(1+h)};$$

where $R_e = \dfrac{(\varepsilon L^4)^{1/3}}{\nu}$ and η is the

dissipative Kolmogorov length.

Integrating over h, the total number of degrees of freedom is,

$$N(R_e) = \int d\rho(h) N_h(R_e) \propto R_e^{\delta};$$

where $\delta = \max_h [d(h)/1 + h]$.

The estimate $l_{min} = \eta(h_{min})$ assures that all the sets $S(h)$ are taken into account. The number of equations which allows us to get such a fully accurate description is thus;

$$N_T^* \sim \left(\frac{L}{l_{min}}\right)^3 \propto R_e^{3/(1+h_{min})};$$

which may be obtained by considering flows in the required number of directions or dimensions.

Directivity:

The directivity inside a given system, which is equal to the tendency of energy transport concerned, shows up as the 'direction' part of the energy quanta that carry out or tend to carry out the transaction. A particle in an isolated box will tend to move in all possible directions. A bias towards any given direction indicates an imbalance of support towards its movement in that given direction and resistance against it. Considering the movement of an energy quantum in a particular direction, this difference between the concerned support and the concerned resistance must be at least the same of the given energy quantum, in accordance to the Theory of Abstraction. For a given quantum-state $(h\upsilon)$,

$$S \sim R$$
$$= h\upsilon \qquad \qquad \dots(32)$$

where, S and R represent the support and resistance respectively.

This means that at least one half of a total energy-quantum gives it its direction while the other part gives it its magnitude. The direction part remains 'hidden' while only the magnitude part shows up as the value of the given quantum state. Considering the direction part however may reduce quantum-transport to classical transport as we shall see here.

For a given transport of energy-quantum, between an initial and a final point, let the trajectory of the initial point $x_o = x(o)$ be denoted by,

$$x(t) = f^t(x_o)$$

Expanding $f^t(x_o + \delta x_o)$ to linear order, the evolution of the distance to a neighbouring trajectory $x_i(t) + \delta x_i(t)$ is given by the Jacobian matrix J,

$$\delta x_i(t) = \sum_{j=1}^{d} J^t(x_o)_{ij} \; \delta x_{oj},$$

$$J^t(x_o)_{ij}$$

$$= \frac{\delta x_i(t)}{\delta x_{oj}} \qquad \qquad \ldots(33)$$

A trajectory of an energy-quantum as moving on a flat surface is specified by two position coordinates and the direction of motion. The Jacobian matrix describes the deformation of an infinitesimal neighborhood of $x(t)$ along the transport. Its eigenvectors and eigenvalues give the directions and the corresponding rates of expansion or contraction. The trajectories that start out in an infinitesimal neighborhood separate along the unstable directions (those whose eigenvalues are greater than unity in magnitude), approach each other along the stable directions (those whose eigenvalues are less than unity in magnitude), and maintain their distance along the marginal directions (those whose eigenvalues equal unity in magnitude).

Holding the hyperbolicity assumption (i.e., for large n the prefactors a_i, reflecting the overall size of the system, are overwhelmed by the exponential growth of the unstable eigenvalues Λ_i, and may thus be neglected), to be justified, we may replace the magnitude of the area of the ith strip $|B_i|$ by $\dfrac{1}{|\Lambda_i|}$ and consider the sum,

$$\lceil n = \sum_{i}^{n} \frac{1}{|\Lambda_i|};$$

where the sum goes over all periodic points of period n. We now define a generating function for sums over all periodic orbits of all lengths,

$$\lceil z = \sum_{n=1}^{\infty} \lceil n \, z^n \qquad \dots (34)$$

For large n, the nth level sum tends to the limit $\lceil n \to e^{-n\gamma}$, so the escape rate γ is determined by the smallest $z = e^{\gamma}$ for which equation (34) diverges,

$$\lceil z \approx \sum_{n=1}^{\infty} (ze^{-\gamma})^n$$
$$= \frac{ze^{-\gamma}}{1 - ze^{-\gamma}} \qquad \dots (35)$$

Making an analogy to the Riemann zeta-function, for periodic orbit cycles,

$$\lceil z = -z \frac{d}{dx} \sum_{p} \ln(1 - t_p);$$

$\lceil(z)$ is a logarithmic derivative of the infinite product

$$\frac{1}{\zeta(z)} = \prod_p (1 - t_p), t_p$$

$$= \frac{z^{n_p}}{|\Lambda_p|} \qquad \text{...} (36)$$

This represents the dynamical zeta function for the escape rate of the trajectories of quantum-transport.

Structure of Matter:

Abstraction says that there is no existence of anything without energy. It also says that points inside energy cluster to form matter of a given property, at the desired scaling-ratio. Let us consider one such system of energy, inside which its constituent points have the tendency to form clusters. In such transactions, the family of evolution-maps f^t form a group. The evolution rule f^t is a family of mappings of strips of transport B, that we may consider, such that,

1) $f^0(x) = x$

2) $f^t[f^{t'}(x)] = f^{t+t'}(x)$

3) $(x, t) \rightarrow f^t(x)$ from $B \times R$ into B is continuous;

where t represents a time interval and $t \in R$.

For infinitesimal times, we may write the trajectory of a given transaction as,

$$x(t + \tau) = f^{t+\tau}(x_0)$$

$$= f[f(x_0, t), \tau] \quad \dots (37)$$

The time derivative of this trajectory at point $x(t)$ is,

$$\frac{dx}{d\tau}\bigg|_{\tau=0} = \partial_\tau f[f(x_0, t), \tau]\big|_{\tau=0}$$

$$= \dot{x}(t) \quad \dots (38)$$

The vector field is a generalized velocity field,

$$\dot{x}(t) = v(x)$$

If x_q represents an equilibrium point, the trajectory remains stuck at x_q forever. Otherwise, the trajectory passing through x_0 at time $t = 0$ may be obtained by,

$$x(t) = f^t(x_0)$$
$$= x_0 + \int_0^t d\tau \, v[x(\tau)], x(0)$$
$$= x_0 \qquad \dots (39)$$

The Euler integrator, which advances the trajectory by $\delta\tau \times \text{velocity}$ at each time step is,

$$x_i = x_i + v_i(x)\delta\tau.$$

This may be used to integrate the equations of the dynamics concerned.

Abstraction in Theory–Zero Postulation Results

Hamiltonian Chaotic Dynamics:

For a Hamiltonian $H(q, p)$ and equations of motion

$$\dot{q}_i = \frac{\partial H}{\partial p_i}, \dot{p}_i = \frac{\partial H}{\partial q_i}$$

With D degrees of freedom, $x = (q, p)$,

$$q = (q_1, q_2, q_3, \dots, q_D),$$

$$p = (p_1, p_2, p_3, \dots, p_D).$$

The value of the Hamiltonian function at the state space point $x = (q, p)$ is constant along the trajectory $x(t)$. Thus the energy along the trajectory $x(t)$ is constant,

$$\frac{d}{dt} H[q(t), p(t)]$$

$$= \frac{\partial H}{\partial q_i} \dot{q}_i(t) + \frac{\partial H}{\partial p_i} \dot{p}_i(t)$$

$$= \frac{\partial H}{\partial q_i} \frac{\partial H}{\partial p_i} - \frac{\partial H}{\partial p_i} \frac{\partial H}{\partial q_i} = 0$$

The trajectories therefore lie on surfaces of constant energy or level sets of the Hamiltonian $[(q, p) : H(q, p) = E]$.

Given a smooth function $g(x)$, the standard map   is,

$$x_{n+1} = x_n + y_{n+1}$$

$$y_{n+1} = y_n + g(x_n).$$

This is an area-preserving map. The corresponding n^{th} iterate Jacobian matrix is,

$$M^n(x_0, y_0)$$
$$= \prod_{K=n}^{1} \begin{pmatrix} 1 + g'(x_K) & 1 \\ g'(x_K) & 1 \end{pmatrix} \qquad \dots (40)$$

Let $M = 1$ as the map preserves areas and also B is symplectic, which in turn is because during cluster formation, the support and resistance between the points tend to attain equillibrium.

The standard map corresponds to the choice $g(x) = k/(2\pi \sin(2\pi x))$. When $k = 0, y_{n+1} = y_n = y_0$, so that angular momentum is conserved, and the angle x rotates with uniform velocity,

$$x_{n+1} = x_n + y_0$$
$$= x_0$$
$$+ (n$$
$$+ 1)y_0$$

The standard map provides a stroboscopic view of the flow generated by a time dependent Hamiltonian,

$$H(x, y, t)$$
$$= \frac{1}{2}y^2$$
$$+ G(x)\delta_1(t) \qquad \ldots (41);$$

where δ_1 denotes the periodic delta function,

$$\delta_1(t) = \sum_{m=-\infty}^{\infty} \delta(t - m)$$

and

$$G'(x)$$
$$= -g(x)$$

A complete description of the dynamics for arbitrary values of the nonlinear parameter k is fairly complex. When K is sufficiently large, single trajectories wander erratically on a large fraction of the phase space.

The component of the dynamics along the continuous symmetry directions of the trajectory behavior or 'drift' may be induced by the symmetries themselves. In presence of a continuous symmetry, an orbit explores the manifold swept by combined actions of the dynamics and the symmetry induced drifts. A group member can be parameterized by angle θ, with the group multiplication law

$$g(\theta') \, g(\theta) = g(\theta' + \theta) \quad \text{and} \quad \text{its}$$

action on smooth periodic functions $u(\theta + 2\Pi) = u(\theta)$ generated by,

$$g(\theta') = e^{\theta' T}, T$$
$$= \frac{d}{d\theta} \qquad \qquad \text{...} (42)$$

The differential operator T is reminiscent of the generator of spatial translations. The constant velocity field $v(x) = v = C.T$ acts on x_j by replacing it by the velocity vector C_j.

Let, G be a group and $gB \to B$ a group action on the state space B. The $[d \times d]$ matrices g acting on vectors in the d-

dimensional state space B from a linear representation of the group G. If the action of every element g of a group G commutes with the flow,

$$g\, v(x) = v(gx), gf^t(x)$$
$$= f^t(gx) \qquad \qquad ...(43)$$

G is a symmetry of the dynamics and is invariant under G or G-equivalent. For any $x \in B$, the group orbit B_x of x is the set of all group actions,

$$B_x$$
$$= [g\, x \mid g \in G]$$

The time evolution and the continuous symmetries can be considered on the same Lie group footing. An element of a compact Lie group continuously connected to identity can be written as,

$$g(\theta) = e^{\theta T}, \theta.T = \sum \theta_a T_a , a$$
$$= 1,2,3, \ldots , N;$$

where $\theta.T$ is a Lie element and θ_a are the parameters of the transformation.

Any representation of a compact Lie group G is fully reducible, and invariant tensors constructed by contractions of T_a are useful for identifying irreducible representations. The simplest such invariant is,

$$T^T.T$$
$$= \sum_\alpha C_2^{(\alpha)} \parallel^{(\alpha)} \qquad \ldots (44)$$

equilibria satisfy $f^t(x) - x = 0$ and relative equilibria satisfy $f^t(x) - g(t)x = 0$ for any t.

A relative periodic orbit is periodic in its mean velocity, $C_p = \theta_p / T_p$ comoving frame, but in the stationary frame its trajectory is quasiperiodic. A relative periodic orbit may be pre-periodic if it is equivariant under a discrete symmetry. Translational symmetry allows for relative equilibria characterized by a fixed profile Eulerian velocity $\mu_{TW}(x)$ moving with constant velocity C, i.e.,

$$\mu(x, t) = \mu_{TW}(x - Ct) \qquad \dots (45)$$

Let, $f\left(\lambda, D\right)$ be a transaction function for a system, where λ is the difference in concentrations of a given observable quantity between two given points of transaction and D the distance between the points. Let, $f_1, f_2, f_3, ..., f_n$ be the complete orthonormal set of eigenfunctions for an operator $\hat{O}$ corresponding to some observable quantity in the system. f can be expanded such that,

$$f$$
$$= j_1 f_1 + j_2 f_2$$
$$+ j_3 f_3 + ... + j_n f_n \qquad ...(46)$$

where $j_1, j_2, j_3, ..., j_n$ are constants.

Operated with $\hat{O}$ from the left on both sides yield,

$$\hat{O}f = j_1\hat{O}f_1 + j_2\hat{O}f_2 + j_3\hat{O}f_3 + \ldots + j_n\hat{O}f_n$$

$f_1, f_2, f_3, \ldots, f_n$ being eigenfunctions of $\hat{O}$, we can write,

$$\hat{O}f = j_1 k_1 f_1 + j_2 k_2 f_2 + j_3 k_3 f_3 + \ldots + j_n k_n f_n$$

where $k_1, k_2, k_3, \ldots, k_n$ are the eigenvalues corresponding to the eigenfunctions.

Considering the complex conjugate of the transaction-function f in equation (46), we have,

$$f^*$$
$$= j_1^* f_1^* + j_2^* f_2^*$$
$$+ j_3^* f_3^* + \ldots + j_n^* f_n^* \qquad \ldots (47)$$

Using these equations, after integrating over all co-ordinate space, we get,

$$\int f^* \hat{O} f \, dt$$

$$= j_1^* j_1 k_1 \int f_1^* f_1 \, dt$$

$$+ j_2^* j_2 k_2 \int f_2^* f_2 \, dt + \ldots + j_n^* j_n k_n \int f_n^* f_n \, dt \quad \ldots (48)$$

In this equation, we have got rid of all the terms of the type $j_a^* j_b \hat{O}_b \int f_a^* f_b \, dt$, as these are all zero because of the orthogonality of the eigenfunctions. Only when $a = b$, are all the terms non-zero and these are the ones we have retained. The integrals on the right side of equation (48) are each equal to one because of the normality condition. Therefore, we write,

$$\int f^* \hat{O} f \, dt$$
$$= j_1{}^* j_1 k_1$$
$$+ j_2{}^* j_2 k_2 + \ldots + j_n{}^* j_n k_n$$

When the system is in the state f, the average value $(\bar{a})$ of the observable k is given by the right-hand side of the previous equation, such that,

$$\bar{a}$$
$$= \int f^* \hat{O} f \, dt \qquad \ldots (49)$$

Relation of group theory to physical

transactions in symmetrical systems:

Say a given dynamic system has a given set of symmetries or stability points. For all points having similar intrinsic properties within such a system, the probability densities of occurrence are equal and must remain unaltered, being all in a similar environment. Thus the energy and Hamiltonian for the system must not change. If E_i is the energy corresponding to the eigenfunction f_i, we may write,

$$\hat{H}f_i = E_i f_i$$

If a symmetry operation ($\hat{x}$) is performed on the system, we have,

$$\hat{x}\,\hat{H}f_i = \hat{x}\,E_i f_i$$

But since $\hat{x}$ does not affect $\hat{H}$ or E, we may write,

$$\hat{H}\,(\hat{x}f_i)$$
$$= E_i\,(\hat{x}f_i) \qquad \ldots (50).$$

The function $\hat{x}\,f_i$ is therefore an eigenfunction of $\hat{H}$ with the same eigenvalues as f_i. We can therefore conclude, if the state is non-degenerate, for normalized functions,

$$\hat{x}f_i$$
$$= \pm f_i$$

Also, as

$$\frac{F}{T}$$
$$= 2c\left(\frac{\lambda}{D}\right)$$

Using Lyapunov exponents for a transport as described by this equation, and replacing $2c\left(\dfrac{\lambda}{D}\right)$ by a quantity τ, we have,

$$\frac{d}{d\tau} f^n(\tau) = \frac{\delta n}{\delta o}$$

i.e.,

$$\frac{\delta n}{\delta o} = \prod_{i=1}^{n} f'(\tau_i) \qquad \ldots (51)$$

$$b = \frac{1}{n} \log_e \left(\frac{\delta n}{\delta o}\right)$$

i.e.,

$$b = \frac{1}{n} \sum_{i=1}^{n-1} \log_e \left| f'(\tau_i) \right| \qquad \ldots (52);$$

where b is a constant (the local slope of all possible routes), and

$$\Psi = \lim_{n \to \infty} \frac{1}{n} \sum_{i=0}^{n-1} \log_e \left| f'(\tau_i) \right| \qquad \ldots (53);$$

where Ψ is a constant.

Using equation (53) the probabilistic plots of all possible routes of transport, i.e., all possible orientations of the points inside the given system of energy (and also the tendencies of these points towards reaching such possible orientations) are to be found for a given scenario. These plots, in turn, yield the description of the structures inside the given system of energy and as such, the description of the structures of matter inside it.

The free-energy change, ΔF_v is obtained from the winding-number W distribution,

$$e^{-\beta \Delta F_v} = \frac{\int \rho_v \left(\Psi, \Psi'; \beta\right) d\Psi}{\int \rho_{v=0} \left(\Psi, \Psi'; \beta\right) d\Psi}$$

$$= \langle e^{i\left(\frac{m}{\tau_i}\right)v.WL} \rangle \qquad \ldots (54).$$

This free-energy change is the Fourier transform of the winding-number distribution and is periodic under

$$v \to v + \frac{h}{mL}$$

Between two open paths m_1 and $m_1 + m$ a single point momentum distribution n_m is the Fourier transform of the off-diagonal density matrix

$$n\left(m_j\right) =$$

$$\frac{\int \rho_b\left(m_1, m_2, m_3, \ldots m_j, m_1 + m, \ldots, m_j; \beta\right) dm_1 \ldots dm_j}{\int \rho_b\left(m_1, m_2, m_3, \ldots m_j, m_1, \ldots, m_j; \beta\right) dm_1 \ldots dm_j}$$

All statistical-mechanical properties of the structures in energy, which in turn are many-point systems themselves, may be determined from the density matrix. The summation over the energy eigenstates of the system;

$$\rho\left(\Psi, \Psi'; \tau_i\right) = \langle \Psi | e^{\beta H} | \Psi' \rangle$$

$$= \sum_j e^{-\beta E_j} \Psi_j\left(\Psi\right) \Psi_j\left(\Psi'\right) \qquad \ldots (56).$$

The identity for the discrete path-integral computations of the density matrix is,

$$\rho\left(\Psi, \Psi'; \tau_i\right)$$

$$= \int (\Psi, \Psi_1; \tau) \rho(\Psi_1, \Psi_1; \tau) \ldots \rho\left(\Psi_{j-1}, \Psi'; \tau\right)$$

$$\times d\Psi_1 \ldots d\Psi_{j-1} \qquad \ldots (57).$$

The amount of information required to describe the trajectory of a concerned plot to within an accuracy or length of measuring tool (ε) be (I_ε), say. We have,

$$I_\varepsilon = \sum_{i=1}^{N} P_i \log_2\left(\frac{1}{P_i}\right);$$

where P_i represents the concerned relative frequencies or probabilities of individual observations. Writing logarithmically (arbitrarily), we have,

$$I_\varepsilon = \mathrm{a} + D_i \log_2\left(\frac{1}{\varepsilon}\right);$$

where a is   constant,   and

$$D_i = \lim_{\varepsilon \to 0} \left[\frac{I_\varepsilon}{\log_2(u/\varepsilon)} - \frac{a}{\log_2(u/\varepsilon)} \right]$$

i.e.,

$$D_i = \lim_{\varepsilon \to 0} \frac{I_\varepsilon}{\log_2(u/\varepsilon)} \ ;$$

$\dfrac{a}{\log_2(u/\varepsilon)}$ being sufficently small to be

neglected (u represents the unit length of the

original).The necessity of I_ε to fall within the

desired value is absolute, barring which safety

of predictions using the plot concerned is

hampered. Ruler-length decreasing to 0, we

have,

$$D_e = \lim_{\varepsilon \to 0} \frac{\log N}{\log_2(1/\varepsilon)};$$

where D_e is the concerned dimension of measurements.

Starting measurements relating to two regions, one initial and the other final, each such region may be further considered to be a collection of some other regions. The scaling operation is performed such that:

1. There is a finite number of sub-divisions.

2. Step 1 is repeated on each new facsimilie.

Thus, we have the dimension,

$$D = \frac{\log N}{\log_2 \left(\frac{1}{r}\right)};$$

where N is the number of facsimilies and r represents the scaling-ratio (i.e., 1/number of sub-divisions).

Also, for a unit length (u), $u = r^D N$

For irregular forms, however, the estimated unit length,

$$L = \varepsilon N$$

Inserting a constant of proportionality (a) in $N = \left(\frac{1}{r^D}\right)$, for going from unit length to a measured length, we get,

$$N = a \left(\frac{1}{r^D}\right)$$

Further, inserting new scaling length (ε) in place of (r), we have,

$$N = a\left(\frac{1}{\varepsilon^D}\right)$$

Thus, from the equations $(L = \varepsilon N)$ and $(N = a/\varepsilon^D)$, we have,

$$L\varepsilon = a\varepsilon^{1-D}$$

When the distance between the concerned initial and final points of transport become sufficiently large, however, than the 'size' of the points, the trajectories tend to become chaotic, thereby increasing the uncertainty of predictions. This uncertainty, in turn, depends upon the Lyapunov exponents (ν). Moreover, there is a stretching or shrinking of a given direction according to the factor $e^{\nu t}$, according as $\mathcal{V}$ being positive or negative in that direction.

Let us suppose a system is characterized by a positive $\mathcal{V}$, i.e., $\mathcal{V}_+$ and its initial state is defined within a size $\mathcal{E}$. Then, in time T, the uncertainty in the co-ordinates concerned will have expanded to the size L of the attractor,

$$L \sim \varepsilon e^{\mathcal{V}_+ T}$$

or,

$$L \sim \varepsilon e^{KT}$$

Either of these relations may be solved for the prediction-time,

$$T \sim \left(\frac{1}{\mathcal{V}_+}\right) \log_e \left(\frac{L}{\varepsilon}\right)$$

or,

$$T \sim \left(\frac{1}{K}\right) \log_e \left(\frac{L}{\varepsilon}\right) \qquad \qquad \dots (58).$$

Conclusion:

We investigate the blueprint for the structures in energy. This helps us to understand the formation of structures of matter as we know them. Abstraction takes into account the scaling-ratio concerned for a given accuracy of observations. At different scaling-ratios, these structures can show completely different emergent phenomena. Again, these emergent phenomena depend upon the given set of dimensions that we consider for a given set of observations. The points inside a given system of energy organize themselves at different scaling-ratios to form different orientations or patterns. These different orientations or patterns may give rise to different orientations of matter.

References:

1. Abstraction In Theory - Laws Of Physical Transactions. Subhajit Ganguly. figshare. http://dx.doi.org/10.6084/m9.figshare.91660 Retrieved 17:31, Jan 26, 2013 (GMT).

2. Condensation States and Landscaping with the Theory of Abstraction. Subhajit Ganguly. figshare.
http://dx.doi.org/10.6084/m9.figshare.91658 Retrieved 17:32, Jan 26, 2013 (GMT).

3. Analysis of the Theory of Abstraction. Subhajit Ganguly. figshare. http://dx.doi.org/10.6084/m9.figshare.91657 Retrieved 17:33, Jan 26, 2013 (GMT).

4. Essentials Of The Theory Of Abstraction - Lecture. Subhajit Ganguly. figshare.

http://dx.doi.org/10.6084/m9.figshare.95936
Retrieved 17:33, Jan 26, 2013 (GMT).

5. Hamiltonian Dynamics in the Theory of Abstraction. Subhajit Ganguly. figshare. http://dx.doi.org/10.6084/m9.figshare.91659
Retrieved 17:34, Jan 26, 2013 (GMT).

6. A Few Implications of the Laws of Transactions, from the Abstraction Theory.. Subhajit Ganguly. figshare. http://dx.doi.org/10.6084/m9.figshare.91655
Retrieved 17:35, Jan 26, 2013 (GMT).

7. Abstraction Theory Central. Subhajit Ganguly. figshare. http://dx.doi.org/10.6084/m9.figshare.91656
Retrieved 17:35, Jan 26, 2013 (GMT).

8. Chaotic Dynamics, Gollub, J.P. Baker, Cambridge University Press.

9. Lyapunov Exponents without Rescaling and Reorthogonization, Salman Habib, Robert D. Ryne, (1995), Physical Review Letters 74: 70-73.

10. Simple Mathematicals models with very complicated dynamics, R.M. May (1976), Nature 261-459.

11. Abstraction In Theory : Laws of Physical Transaction, Subhajit Ganguly, VMA Publications and Createspace, 2012.

12. Ali H. Chamseddine1,2 and Alain Connes2, A Universal Action Formula, arXiv:hep-th/9606056v1

13. A. Connes, Publ. Math. IHES 62, 44 (1983); Noncommutative Geometry (Academic Press, New York 1994).

14. A. Connes and J. Lott, Nucl. Phys. Proc. Supp. B18, 295 (1990); proceedings of 1991

Carg`ese Summer Conference, edited by J. Fröhlich et al (Plenum, New York 1992).

15. D. Kastler, Rev. Math. Phys. 5, 477 (1993).

16. Yasuhiro OKADA, Masahiro Y AMAGUCHI*) and Tsutomu YANAGIDA, Upper Bound of the Lightest Higgs Boson Mass in the Minimal Supersymmetric Standard Model, Prog. Theor. Phys. Vol. 85, No.1, January 1991, Progress Letters.

17. L. Girardello and M. T. Grisaru, Nucl. Phys. B194 (1982), 65.

18. K. Inoue, A. Kakuto, H. Komatsu and S. Takeshita, Prog. Theor. Phys. 68 (1982), 927.

19. L. Alvarez-Gaume, J. Polchinski and M. B. Wise, Nucl. Phys. B221 (1983), 495.

20. J. Ellis, J. S. "Hagelin, D. V. Nanopoulos and K. Tamvakis, Phys. Lett. 125B (1983), 275.

21. L. Ibanez and C. Lopez, Phys. Lett. 126B (1983), 54.

22. H. P. Nilles, Phys. Rep. 110 (1984)

23. S. P. Li and M. Sher, Phys. Lett. 140B (1984), 339.

24. M. S. Berger, Phys. Rev. D41 (1990), 225.

25. J. F. Gunion and A. Turski, Phys. Rev. D39 (1989), 2701; D40 (1989), 2325, 2333.

26. Y. Okada, M. Yamaguchi and T. Yanagida, Tohoku University Preprint, TU-363.

27. Walter Pfeifer, An Introduction to the Interacting Boson Model of the Atomic Nucleus, vdf Hochschulverlag.

28. M. Veltman, Acta. Phys. Po!. B12 (1981), 437.

29. L. Maiani, Proc. Summer School of Gil-Sur-

Yvette (Paris, 1980), p. 3.

30. E. Witten, Nucl. Phys. B188 (1981), 513.

31. M. Dine, W. Fischler and M. Srednicki, Nucl. Phys. B189 (1981), 575.

32. K. Inoue, A. Kakuto, H. Komatsu and S. Takeshita, Prog. Theor. Phys. 67 (1982), 1889.

33. R. Flores and M. Sher, Ann. of Phys. 148 (1983), 95.

34. Z Physics at LEP, CERN Yellow Book, ed. G. Altarelli, R. Kleiss and C. Verzengnassi (CERN, Geneva, 1989).

35. J. Kalinowski, B. Grzadkowski and S. Pokorski, Phys. Lett. B241 (1990), 534.

36. R. M. Barnett and G. Gamberini, Phys. Lett. B241 (1990), 541.

37. A. Djouadi, J. L. Kneur and G. Moultaka,

Phys. Lett. B242 (1990), 265.

38. Z. Kunszt and W. J. Stirling, Phys. Lett. B242 (1990), 507.

39. E. Fradkin and A. Tseytlin, Nucl. Phys.

40. K.S. Stelle, Phys. Rev. D16, 953 (1977).

41. E. Tomboulis, Phys. Lett. 70B, 361 (1977).

42. For a review see G. Ross, Grand unified theories, Frontiers in Physics Series, Vol.60 (Benjamin, New York).

43. M. B'eg, C. Panagiotakopoulos and A. Sirlin, Phys. Rev. Lett. 52, 883 (1984);

44. M. Lindner, Z. Phys. C31, 295 (1986).

45. Felipe Cucker, Steve Smale, On the mathematics of emergence, Japanese Journal of Mathematics. 197–227 (2007)

46. ALBERTO S. CATTANEO AND GIOVANNI FELDER, A PATH INTEGRAL APPROACH TO THE KONTSEVICH QUANTIZATION FORMULA, http://arxiv.org/abs/math/9902090v3

47. Bruce J. Berne, ON THE SIMULATION OF QUANTUM SYSTEMS" PATH INTEGRAL METHODS, Annu. Rev. Phys. Chem. 1986.37:401-424.

ABOUT THE AUTHOR

Subhajit Ganguly is a physicist whose areas of expertise include the Theory of Abstraction. His contribution to the theory is noteworthy, to say the least. The other areas of science that he has made notable contributions to include astronomy, mathematics and the Chaos Theory. Zero-postulation is a new concept he has introduced to the theorizing process in sciences. He is an Advisor to the Figshare Open Science Platform and is an Ambassador for the Open Knowledge Foundation.